SCIENTIFIC AMERICAN EDUCATIONAL PUBLISHING

IN MOTION

10 FUN EXPERIMENTS EXPLORING FORCES

BRING SCIENCE HOME

Published in 2024 by Scientific American Educational Publishing in association with
The Rosen Publishing Group.

Contains material from Scientific American, a division of Springer Nature America, Inc., reprinted by permission as well as original material from The Rosen Publishing Group.

First Edition

Editor: Kristen Rajczak Nelson
Designer: Rachel Rising

Activity on page 5 by Science Buddies (November 12, 2015); page 11 by Science Buddies (July 9, 2015); page 15 by Science Buddies (August 23, 2012); page 21 by Science Buddies (July 12, 2012); page 27 by Science Buddies/Ben Finio (October 1, 2015); page 31 by Science Buddies (January 19, 2012); page 37 by Science Buddies (November 6, 2014); page 43 by Science Buddies (August 9, 2012); page 49 by Science Buddies (August 2, 2012); page 55 by Exploratorium (November 20, 2014).

All illustrations by Continuum Content Solutions

Photo Credits: pp. 3, 5, 8, 11, 13, 14, 15, 18, 19, 21, 25, 30, 31, 34, 40, 41, 46, 49, 52, 53, 58 cve iv/Shutterstock.com; pp. 4, 5, 11, 15, 21, 27, 31, 37, 43, 49, 55 Anna Frajtova/Shutterstock.com.

Library of Congress Cataloging-in-Publication Data

Names: Scientific American, inc., editor.
Title: In motion : 10 fun experiments exploring forces / Scientific American Editors.
Other titles: 10 fun experiments exploring forces
Description: Buffalo, NY : Scientific American Educational Publishing, an imprint of Rosen Publishing, [2024] | Series: Bring science home | Includes bibliographical references and index.
Identifiers: LCCN 2022061451 | ISBN 9781725349162 (library binding) | ISBN 9781725349155 (paperback) | ISBN 9781725349179 (ebook)
Subjects: LCSH: Force and energy--Experiments--Juvenile literature. | Science projects--Juvenile literature.
Classification: LCC QC73.4 .I56 2024 | DDC 531/.6078--dc23/eng20230323
LC record available at https://lccn.loc.gov/2022061451

Manufactured in the United States of America

CONTENTS

Introduction 4

Build a Paper Rocket 5

Make a Whirlybird from Paper* 11

Stability Science: How Tails Help a Kite Fly* 15

Spinning Science: Centripetal Force Using Marbles in Jell-O* 21

Spinning Symmetry with Pinwheels 27

Fight Slippage with Friction* 31

Speedy Science: How Does Acceleration Affect Distance? 37

Soaring Science: The Aerodynamics of Flying a Frisbee 43

Suspended Science: How Does a Hovercraft Hover?* 49

Loop-the-Loop with a Flying Hoopster 55

The Scientific Method 61

Glossary 62

Additional Resources 63

Index 64

* THESE ACTIVITIES INCLUDE SCIENCE FAIR PROJECT IDEAS.

INTRODUCTION

Doing science experiments in the classroom helps you further understand a variety of concepts, but the ability to complete projects in your own home brings comprehension to a new level! Concepts such as force, friction, pressure, velocity, and more are utilized in the following experiments. Get ready to create science projects out of accessible household items and better understand the world around you!

Projects marked with ⚛ include a section called Science Fair Project ideas. These ideas can help you develop your own original science fair project. Science fair judges tend to reward creative thought and imagination, and it helps if you are really interested in your project. You will also need to follow the scientific method. See page 61 for more information about that.

Build a Paper Rocket

MAKE A ROCKET THAT SOARS—WITH SCIENCE! LEARN ABOUT THE IMPORTANCE OF STABILITY IN FLIGHT, AND SEE HOW FAR YOUR PAPER ROCKET CAN FLY.

Blast off! Have you ever played with a model or toy rocket, or seen a real rocket launch on TV? In this project, you will make simple rockets out of paper and launch them by blowing into a drinking straw. Can you make the rocket that flies the farthest?

PROJECT TIME

30 to 40 minutes

KEY CONCEPTS

Physics
Aerodynamics
Flight
Stability

BACKGROUND

All flying objects, from rockets to airplanes to birds, have something in common—they need to remain *stable* when they fly. You are probably pretty familiar with what "stability" means for objects on the ground. Did you use training wheels when you learned how to ride a bike? Training wheels help keep the bike stable so you do not fall over. The same concept applies to things that fly. They need to stay pointed in the same direction when they fly forward, without spinning or tumbling, which could cause them to crash.

You may have noticed that rockets and missiles usually have triangular fins at their bases. The same applies to other long, skinny objects that fly through the air quickly, such as arrows. What purpose do these fins serve? Can they help the rockets fly better? Try this project to find out!

MATERIALS

- Two pieces of paper
- Scissors
- Pencil
- Drinking straw
- Scotch tape
- Ruler
- Clear space in which to launch your "rockets," such as a large room, hallway or outdoor area with no wind
- Measuring tape (optional)

PREPARATION

- Gather all of the materials you will need for the activity.

PROCEDURE

- Cut one piece of paper into four smaller rectangles, by cutting it in half lengthwise and widthwise. This will allow you to make four rockets.

- Wrap one of the paper rectangles around a pencil to form a cylinder, with the long edge of the paper along the length of the pencil.
- Tape the cylinder closed so it does not unravel (but do not tape it to the pencil).
- Slide the cylinder off the pencil. Pinch one end of the cylinder shut and seal it with tape. (This is the "front" end of your rocket.) Leave the other end open. This will be your first rocket, with no fins.
- With plenty of room in front of you—and no obstructions, such as furniture or people—prepare to launch your first rocket! Slide it over a drinking straw. Aim the straw forward, then blow into it as hard as you can. Watch your rocket as it flies. *How far does it go? Does it fly straight or does it tumble in midair?*
- Launch your rocket a few more times to see if it flies the same way. If you would like to record your rocket flight distances, be sure to launch it from the same place each time, and measure to the landing spot with a tape measure.
- Make another paper rocket following the previous steps. Remember to pinch one end and tape it shut.
- For this rocket, however, you will make fins. Cut out two right triangles (with a 90-degree angle in one corner) from the other piece of paper. The long sides of the triangles should be about 3 inches (8 cm). You will fold each triangle to make two fins, so you will have four fins total.
- Draw a line that splits one triangle in half (from the 90-degree corner to the middle of the long side of the triangle).
- Draw two lines parallel to the first line (one on each side), about 0.2 inch (5 mm) away from it.
- Now, fold the triangle up along these two lines. The result should be two triangles sticking up in the air (the fins), with a flat part connecting them in between.

- Tape the flat part to the side of your cylinder, toward the open end (the base, or bottom, of your rocket).
- Repeat these steps for the other triangle, and tape it to your cylinder on the opposite side of the first one. The result should be four fins that form a "+" shape when you look at the rocket from either end. If necessary, bend the fins so they are spaced out 90 degrees apart from one another.
- Slide the new rocket onto the drinking straw and launch it. *How far does this rocket go? How does its flight compare with your first finless rocket? Does it go farther? Does it tumble or does it fly straight? Do you think fins help the stability of your rocket?*
- Launch it a few more times. If you are measuring the flight distance of each rocket, use a tape measure and record how far it flew.

EXTRA

Try different numbers of fins. *For example, what happens if you only use two fins instead of four?*

EXTRA

Try different shapes for fins. *For example, what happens if you make semicircular fins instead of triangles?*

EXTRA

Try attaching the fins at different points along the length of your rocket. *Do the fins still work if you put them in the middle or front of the rocket instead of the back?*

OBSERVATIONS AND RESULTS

You should have seen that your finless rocket flew straight at first but quickly spiraled out of control. It might have tumbled through the air and fluttered to the ground, almost like a leaf falling from a tree. This is because the rocket did not have fins to keep it stable. If it started turning just a little bit, then it would start turning even more rapidly until it completely lost control. In contrast, your second rocket that had fins should have flown straight, and traveled much farther as a result. This is because the fins help keep the rocket stable, or pointed in the same direction. If the rocket turns a little bit, the fins help turn it back in the original direction.

CLEANUP

Recycle your rocket pieces afterwards!

Make a Whirlybird from Paper

FLY WITH PHYSICS! LEARN HOW HELICOPTERS STAY ALOFT BY TRYING THIS FUN SCIENCE ACTIVITY WITH A TWIST!

Have you ever seen a helicopter flying through the air? Have you ever wondered how they fly—or if you could try flying one yourself? This fun activity will help you get started at home building a simple paper helicopter. And you will learn a little bit about what keeps these amazing vehicles aloft.

PROJECT TIME

About 20 minutes

KEY CONCEPTS

Gravity
Weight
Lift
Aerodynamics

BACKGROUND

Helicopters stay in the air using spinning blades that are used to generate lift. With enough of it, a craft can overcome the force of gravity, which pulls the object down toward Earth. Aircraft such as helicopters with spinning blades are called rotary wing, unlike traditional airplanes, which are fixed wing.

In this activity, you will build a simple paper helicopter called a "whirlybird." Unlike a real helicopter, the whirlybird does not have a motor to make its blades spin. Due to its special shape, however, the blades will still spin as it falls. This generates additional lift that slows the whirlybird even as it drops. So, it will fall much slower than if you crumpled up the same piece of paper and dropped it. Do you think adding paper clips as weights to the whirlybird will make it fall faster? Try this activity to find out!

MATERIALS

- Computer with access to a printer to print the whirlybird template (If you do not have access to a printer, you can use a ruler and pencil to draw your own whirlybird template based on the online one.)
- Printer paper
- Scissors
- Several paper clips
- Stopwatch (optional)
- A safe, high place from which to drop the whirlybirds (You could have an adult stand on a chair or stepping stool, for example.)

PREPARATION

- Download and print the whirlybird template from this pdf: **https://www.sciencebuddies.org/science-fair-projects/whirly-bird-template.pdf**. If you do not have access to a printer, you can download the file and open it on your computer, then use a pencil and ruler to draw the whirlybird shape on a piece of paper, based on the dimensions in the template.

- Follow the directions on the template to cut out and fold your whirlybird.

PROCEDURE

- Now it's time to drop your whirlybird! What do you think will happen when it gets dropped? Have an adult help drop it from a safe high place (such as standing on a chair or a stepping stool). *How did the whirlybird fall?*
- Now drop the whirlybird a couple more times—in the same fashion, from the same height. *Why do you think it is important to drop it more than once?*
- Now, attach a paper clip to the bottom part of the whirlybird and drop it again. *Do you think it will fall faster? Pay close attention. How did it fall with a paper clip attached?*
- Keep adding paper clips, one at a time, to your whirlybird and drop it after each new paper clip. *What do you observe as you add more and more paper clips?*
- You can also use a stopwatch to time how long it takes the whirlybird to fall with different numbers of paper clips on it. Write down the times to help you remember how fast it falls when you add different numbers of paper clips. *Do you detect a pattern?*

EXTRA

Try making whirlybirds from different types of paper, such as printer paper, construction paper, and cardstock. *Do the different types of paper fall at different speeds?*

SCIENCE FAIR IDEA

Try modifying the whirlybird template or making your own design. *For example, what happens if you make the "wings" longer or shorter or narrower or wider? What if you make the edges of the wings wavy or zigzagged instead of straight? What if you make the wings curved instead of rectangular?* Try out a bunch of different ideas to see what happens. Then try to design your own whirlybird—one that falls as slowly as possible.

OBSERVATIONS AND RESULTS

When you drop a whirlybird, it will take a fraction of a second for it to start spinning and slow down. This is why you need to drop it from a high location—it still needs a certain amount of vertical space to begin its spinning. Once the whirlybird does start spinning, it should "helicopter" slowly to the ground. As you add paper clips, the whirlybird should fall faster and faster until eventually it drops so fast that it does not spin at all.

Experimenting with your own whirlybird designs can be a fun and challenging engineering activity. There might not be one single design for the "best" whirlybird. Some designs might fall more slowly than others. If you change the dimensions too dramatically, however, a whirlybird may actually become unstable and tumble to the ground instead of spinning.

CLEANUP

Return the office products to their location and recycle the paper pieces.

Stability Science
How Tails Help a Kite Fly

A KITE CAN'T FLY WITHOUT A TAIL, AND IT CERTAINLY CAN'T FLY WITHOUT YOUR GUIDANCE! LEARN HOW TO HELP A KITE FLY THROUGH STABILITY SCIENCE.

Have you ever tried to build your own kite? Kites have been a source of entertainment through the centuries for kids worldwide. And they've been used for scientific experiments, too—Benjamin Franklin flew one to investigate lightning. (Something you shouldn't do!) In this activity, you will have a chance to build your very own kite—a simple sled kite—and use it to investigate how tails help kites fly. After trying out this activity, you may add tails to other kites to help them soar to greater heights!

PROJECT TIME

20 to 30 minutes

KEY CONCEPTS

Aerodynamics
Wind
Flight
Forces

BACKGROUND

Did you know that, besides Ben Franklin, one of the most famous kite flyers of all time was a 10-year-old boy? His name was Homan Walsh, and without him the Niagara Falls Suspension Bridge may not have been built in 1847. Before construction could begin, someone needed to get a line across the Niagara River Gorge. Homan successfully flew a kite from one side of the gorge to the other over the treacherous rapids below and, after securing his kite string, heavier line was fed across until a steel cable could connect both shores and bridge construction could begin. Homan was rewarded with a $10 cash prize, which was a lot of money in 1847!

How does a kite fly? As someone runs with a kite, the wind going head-on into the kite causes a lift force on it. This force is perpendicular to the wind, pushing the kite up. At the same time, the force of drag pulls the kite back, in the direction that the wind is going.

There are many different kite designs. Some are very old, like traditional Chinese and Japanese kites. Some designs are very new, like the dynamic stunt kites used in sport-kite flying competitions. Their modern materials and designs make them ultra-maneuverable.

MATERIALS

- Printer
- Paper
- Scissors
- Crayons and markers (optional)
- Tape
- Two drinking straws
- Hole punch
- Kite string
- Measuring tape or ruler
- Paper clip
- Plastic grocery bag or other thin plastic bag
- An open, clear area outside

PREPARATION

- Download the Sled Kite Template and print it out on a sheet of 8 1/2- by 11-inch (21.5- by 28-cm) paper: **https://www.sciencebuddies.org/science-fair-projects/project_ideas/Aero_kitetemplate.pdf.**
- Carefully cut out the sled kite. You can decorate it using crayons, markers, or other media.
- Trim the length of the two drinking straws so they will fit in the area marked for the straws. Tape the straws in place.
- Place three pieces of tape in the marked areas covering the black circles to reinforce the holes for the kite string. Using a hole punch, carefully punch the two holes marked by the black circles.
- Cut two pieces of kite string to make each 17 inches (45 cm) long. Tie a string through each hole. Tie them tight, but not so much that you tear the paper. Tie the opposite end of both strings together to one end of a paper clip.
- Cut a long piece of kite string (about 3 feet [0.9 m]). Tie one end of this string to the other end of the paper clip. Your sled kite is ready to fly!
- To make some tails to test on your kite, take a plastic grocery bag or other thin plastic bag and lay it completely flat. If the bag has handles, cut with scissors straight across to remove the handles. Then keep cutting the bag in this way to create thin rings (which will look like strips when flat)—each should be about 1 inch (3 cm) wide. Cut up the whole bag into these rings. Discard the handles.

PROCEDURE

- Try to fly your kite without a tail in an open, clear area outside. First walk with it, then try running with it. *What does the kite do when you walk and then when you run? How well does it fly?*

- Using the plastic grocery bag rings you made, tape a tail to the bottom of your kite that is 4 inches (10 cm) long. (This will probably be less than one grocery bag ring.) The tail should be centered. Try to fly your kite outside again, first walking and then running with it. *How does the kite fly compared with when it had no tail? Why do you think this is?*
- You'll now test a tail that is 40 inches (100 cm) long. To make this tail, loop two rings together and gently pull them tight. Attach more rings to the tail this way to make it longer. Attach the longer tail to the 4-inch (10-cm) tail on the kite to get to 40 inches (100 cm) long. Try to fly your kite outside again, first walking and then running with it. *How well does the kite fly with the 40-inch (100-cm) tail compared with the 4-inch (10-cm) one? How does it compare with flying a kite without a tail? Why do you think this is?*
- You'll now test a tail that is 200 inches (508 cm) long. Use the same looping process to expand your 40-inch (100-cm) tail another 160 inches (406 cm) in length. Try to fly your kite outside again, first walking and then running with it. *How well does the kite fly with the 200-inch (508-cm) tail compared with the other tails or no tail? What differences do you notice? Why do you think this is?*
- *Out of the tails you tested, which tail do you think helped the kite fly the best? Why do you think this is?*

EXTRA

Do this activity again, but instead of comparing different tail lengths, try adding additional tails and comparing the results. Add them symmetrically to the back end of the kite. *How well does the kite fly with no tail, one tail, or two tails?*

EXTRA

In this activity, you flew a kite using kite string 3 feet (0.9 m) long. *What happens when you fly the kite using different lengths of kite string, such as 39 inches (1 m), 118 inches (3 m), or 196 inches (5 m)?*

SCIENCE FAIR IDEA

There are a lot of other kite designs. Use the internet, books, or magazines at a library to find plans for different kinds of kites. Then test your new kite designs as you did in this activity. *How does the tail length affect how well other kite designs work? What other variables affect how well kite designs work?*

OBSERVATIONS AND RESULTS

When you tried to fly the kite without a tail, did it spin and roll around a lot, and not really fly at all? Did flying the kite with a 40-inch (100-cm) tail work the best?

Trying to fly a kite without a tail may result in the kite spinning and rolling a lot because the kite is unstable. Adding a tail to a kite helps make it fly more stably by adding some needed weight and drag to its lower end. Adding a small tail, such as the 3-inch (10-cm) tail, may help the kite fly a little more stably, but it will still spin and roll around a lot. Adding a longer tail, such as the 40-inch (100-cm) tail, should help the kite fly well, allowing it to get relatively high without rolling much. A very long tail, such as the 200-inch (508 cm) tail, will help keep the kite from rolling, but the tail may be so heavy that it will prevent the kite from flying very high. Some recommend adding a tail that's around three to eight times the kite's length—but the best way to determine the right tail length for your particular kite is to try it out!

CLEANUP

Recycle any paper pieces and return the office supplies to their location, or keep the kite for future flights!

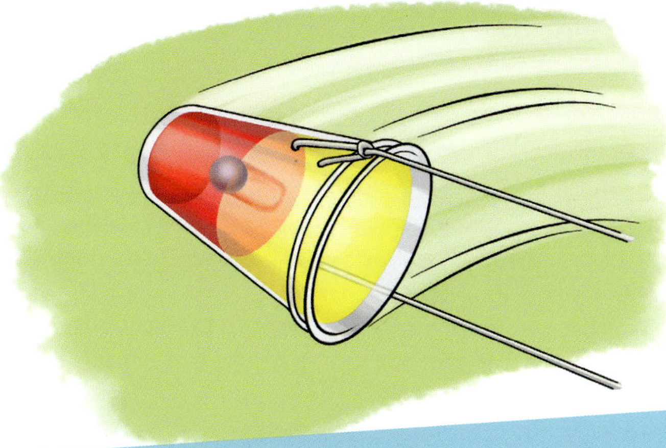

Spinning Science

Centripetal Force Using Marbles in Jell-O

USING CENTRIPETAL FORCE AND JELL-O, DISCOVER THE LAWS OF MOTION IN YOUR OWN HOME!

PROJECT TIME

6 hours

KEY CONCEPTS

Physics
Forces
Motion
Centripetal force
Circular motion
Acceleration

Have you ever wondered what keeps you in your seat when you're riding a giant loop-de-loop roller coaster? Surprisingly, it's not only the seat belt! You're kept in your seat because of something called centripetal force. Centripetal force actually does much more than make a ride on a roller coaster's loop possible—it keeps satellites in orbit and you in your bicycle when taking a turn! In this activity, you will use marbles and Jell-O to investigate centripetal force and circular motion. Which way will the marbles go?

BACKGROUND

Isaac Newton, a careful observer of nature, used mathematics and science to describe natural phenomena that were not understood at the time. Newton's laws of motion are still used today, and these principles can be found in almost any moving technology. Newton discovered that for an object to move, it must be acted on by a force that makes it move in a certain direction. The object set in motion will continue this motion until it meets an opposite force. You have felt this phenomenon when riding in a car. When the car starts moving, you rock backward because your body wants to stay in its stationary position. After you're moving, if the car suddenly stops, you will rock forward because your body wants to keep moving at the same speed and direction.

Newton realized that when an object moves (or accelerates) in a circle, the object will move outward, away from the circle's center. For example, when you're riding in a car and it makes a turn, your body moves to the outside of the turn, away from the turn's direction. The force that makes you follow a circular path, even though your body wants to move away from the turn, is called centripetal force.

MATERIALS

- Transparent plastic cups, at least four
- Scissors or one-hole puncher
- Duct tape or electrical tape
- String
- Cooking pot
- Measuring cup
- Stove
- Jell-O, two packages, each of a different color
- Refrigerator
- Marbles, at least three
- An open area outside
- Flashlight

PREPARATION

- Use the scissors or one-hole puncher to make a small hole about 1 inch (2.5 cm) from the top rim of one of the plastic cups. Make a second hole on the opposite side of the cup.

- Put a small piece of duct tape or electrical tape on the edge of the cup, just above each hole. Fold over the tape so it is on the outside and inside of the cup, but not blocking the holes. This will help prevent the string from detaching.
- Attach the string to the cup, tying one end of the string through one of the holes at the cup's top and the other end to the other hole.
- This cup will be your centripetal force generator. Test to make sure that the string is strongly attached to the cup by holding on to the string and pulling down on the cup.
- Ask an adult to help you make the Jell-O. Be careful when working with the boiling water. Also, be careful not to let the Jell-O spill and stain anything.

PROCEDURE

- Make one of the packets of Jell-O by dissolving it in the appropriate amount of water according to the directions on the package.
- Pour the Jell-O into the three other plastic cups, filling each cup halfway full.
- Place these cups in the refrigerator and chill until the Jell-O is completely set, about two to three hours.
- When the Jell-O is set, in each cup place a marble on the surface of the Jell-O in the center of the cup. Gently press it into the Jell-O just until the marble is secure and will not move around. *Why do you think it is important that the Jell-O is firm enough to support the weight of the marble?*
- Make the second batch of Jell-O. *Why do you think the second batch should be a different color from the first?*

- Slowly and carefully pour the hot Jell-O into the cups, covering the first layer of Jell-O and the marble until the cup is almost full, leaving about 1 inch (2.5 cm) at the top of the cup.
- Place the cups back in the refrigerator until the mixture is completely set, about two to three hours.
- When the Jell-O cups are set, put one of them into the centripetal force generator cup (the cup with a string tied to it) by stacking the cup with Jell-O inside the centripetal force generator cup.
- Take the stacked cups outside to an open area. (If the string on the centripetal force generator cup broke by accident, you would not want to get Jell-O all over the floors, walls, and furniture!)
- Hold the string and twirl the stacked cups around your head for 20 revolutions, counting each time the cups make a complete circle. *If you were to let go of the cups while spinning them, in what direction do you think they would go?*
- After 20 revolutions, stop spinning and remove the inner cup from the outer cup.
- Shine a flashlight through the cup with the marble. *Can you see the marble? Where is it? How has it moved?*
- **Tip:** If you have trouble locating the marble, try backlighting the marble by shining the flashlight through the back of the cup, toward you.
- **Tip:** If the marble did not move, the Jell-O may be too firm. First try spinning it around for 20 revolutions again—but this time spin it harder. If the marble has still not moved, either let the cups sit out at room temperature overnight to soften the Jell-O or repeat this activity, this time using more water to make the Jell-O.

- One at a time, spin the other Jell-O cups in the centripetal force generator cup for 20 revolutions. *Do you notice any patterns of movement? Did the marbles always move in the same direction? Did they move in the direction you thought they would? How far did they move?*

EXTRA

One of Newton's other laws says that there is a relationship between the motion of an object and its mass. Try a similar activity with small objects of different weights to see if this has an effect on the amount of movement an object makes due to centripetal force. Instead of marbles, try lead fishing weights, beans, quarters, beads, etcetera. *Do you notice a difference in motion between objects of different weights or sizes?*

EXTRA

How fast did you swing your centripetal force generator? Is there a relationship between speed and circular motion? Try using a metronome to guide your speed of rotation, setting the metronome at fast and slow speeds. *Do the cups move differently at faster speeds than they do at slower speeds?*

SCIENCE FAIR IDEA

There are many ways to investigate Newtonian motion. Try taking some of your cups containing Jell-O and marbles with you for some on-site experimentation. *What happens to the marble when you take it on a swing, slide, merry-go-round, roller coaster, car ride, bike ride, or anywhere else you are curious to find out about?*

OBSERVATIONS AND RESULTS

In each cup, did the marble move from the middle of the cup to the bottom?

When an object moves, or accelerates, in a circle, the object wants to move out, away from the circle's center. Without the push of centripetal force, the object would move in a line, flying out and away from the center of the circle. For example, if you had let go of the cups when you were spinning them, they would fly away from you in a straight line. Each marble in the spinning cups also wanted to move away from you, away from the circle's center, and so each marble should have traveled through the Jell-O and ended up at the bottom of the cup.

In this activity, centripetal force acting on the system is supplied by the tension in the string. This force kept the cups moving in a circular path. This is a "pull" force, similar to how satellites are kept in orbit due to the pull of Earth's gravity. A "push" centripetal force acted on the marble: this force, supplied by the bottom of the cup, kept the marble going in a circular path and not flying away in a straight line. A push centripetal force also keeps you in your seat on a loop of a loop-de-loop roller-coaster ride and when you're making a turn on your bicycle.

CLEANUP

Throw the Jell-O in the trash when you are done after retrieving your marbles.

Spinning Symmetry with Pinwheels

WHY DO WIND TURBINES AND AIRPLANE PROPELLERS LOOK THE WAY THEY DO? THE SECRET IS IN THE SYMMETRY. LEARN WHY ENGINEERS DESIGN THEM THIS WAY WITH THIS FUN, HANDS-ON ACTIVITY. GIVE IT A WHIRL!

Have you ever played with a toy pinwheel—or seen larger real-life versions, such as a windmill or wind turbine? Have you wondered why they look the way they do—with multiple blades, arranged symmetrically? Why can't they get away with just one blade? Or blades of different shapes and sizes? Try this fun project to find out!

PROJECT TIME

20 minutes

KEY CONCEPTS

Forces
Friction
Symmetry
Wind power

BACKGROUND

Spinning blades are used in a wide variety of machines. Some blades are built to harvest energy from the air, such as on windmills and turbines. The wind pushes on these blades making them spin, and then the energy from their movement is transferred to turn gears or generate electricity. There are also blades that use a powered motor to push on air—sometimes to move it (as with a fan) or to push an object through it (for instance, an airplane propeller).

One thing all of these blades have in common is that they are installed in a symmetric pattern. This means they are equally spaced around the center, or "balanced." There might be two blades directly across from one another or four blades that form an "X" and meet at right angles in the middle. There could even be an odd number of blades (for example, three blades evenly spaced would be at 120-degree angles to one another). Why is it important for the blades to be symmetric? Why don't you see windmills with just one blade or with two blades on the same side? This project will show you the best way to get the blades spinning!

MATERIALS

- At least one pinwheel (that can be cut apart)
- Hair dryer
- Scissors
- Adult supervision

PREPARATION

- Gather all of your materials in an area where you can plug in and use the hair dryer. Have an adult operate the hair dryer about 3 feet (0.9 m) away from where you will be standing with the pinwheel.

PROCEDURE

- Turn on the hair dryer.

- Hold the pinwheel about 3 feet (0.9 m) away from the hair dryer. Gradually move the pinwheel closer to the dryer until it begins spinning. *How close does the pinwheel need to be before it starts spinning?*

- Now, have an adult help you use scissors to cut off one blade from the pinwheel as close to the base as you can. This makes the pinwheel asymmetric because the remaining blades are no longer evenly spaced around the center.

- Again, hold the pinwheel about 3 feet (0.9 m) in front of the hair dryer and gradually move it closer until it starts spinning. *How close does the pinwheel have to be before it starts spinning this time?*

- Use scissors to cut another blade off the pinwheel. Depending on how many blades your pinwheel had initially, you will have to make a decision about which blade to cut and whether the pinwheel remains asymmetric or becomes symmetric again. For example, suppose your pinwheel had four blades to start. When you cut off one blade, it has three unevenly spaced blades, so it is asymmetric. Now you can cut off a second blade next to the one you just cut off, so the pinwheel will still be asymmetric, with two unevenly spaced blades. Or you could cut off a second blade across from the one you just cut off. This will make the pinwheel symmetric again, with two blades directly across from one another. If you have more than one pinwheel, you can try different combinations.

- Hold the pinwheel 3 feet (0.9 m) in front of the hair dryer again and slowly move it closer. *How close does the pinwheel need to be before it starts spinning this time?*

- Repeat this process until your pinwheel only has one blade left.

- *Does the number of blades affect how easily the pinwheel spins? Do you have to hold the pinwheel closer to the hair dryer sometimes?*

- *Does how easily the pinwheel spins depend on whether the blades are symmetric or asymmetric?*

EXTRA

Use a ruler or measuring tape to record how far the pinwheel needs to be from the hair dryer to start spinning in each trial. Then you can make a graph of your results with the distance from the dryer versus the number of blades.

OBSERVATIONS AND RESULTS

You should have seen that it was "easiest" for the pinwheel to spin when it had a larger number of symmetric blades. As you removed blades, you might have needed to hold the pinwheel closer and closer to the hair dryer to make it spin. This is especially true if the blades were asymmetric. This can cause the blades to "wobble" and create more friction on the pinwheel's axle (the rod that supports the blades). Higher friction makes it harder for the blades to spin, meaning the wind must be stronger—so you have to hold the pinwheel even closer to the hair dryer to get it to spin. How do you think this affects engineers' plans when they design and make wind turbines or airplane propellers?

CLEANUP

Return the hair dryer to its proper location with the help of an adult, and discard the rest of the items in the trash.

Fight Slippage with Friction

DISCOVER ONE OF THE MAIN RESISTING FORCES FOUND IN YOUR DAILY LIFE.

PROJECT TIME

20 minutes, plus overnight freezing

KEY CONCEPTS

Physics
Friction
Mass
Gravity

Have you ever driven up a mountain, seen a sign reading "Hazard! Icy Roads Ahead," and wondered why ice makes roads dangerous to drive on? The answer has to do with friction—specifically, the lack of it.In the case of driving a car down the road, the friction that allows the car to move occurs between the tires and the road. This friction normally allows the car to "grip" the road—keeping the tires in contact with the relatively rough road surface, and the driver of the car in control. However, when the road is icy, the friction between the two surfaces decreases, allowing the car to slip—rather than grip. You can replicate these situations much more safely at home using some common objects—regardless of the weather.

BACKGROUND

Friction is the resistance to motion when two objects rub together. An example of low friction is when you sled down an icy hill. It's easy because both the sled and the ice on the hill are smooth and the surfaces of the sled's runners and the ice slide past one another with little resistance. On the other hand, sledding down a rough, concrete driveway is almost impossible due to the rough surface of the driveway, which provides resistance on the sled's runners. This is an example of high friction.

As you can imagine, friction happens when the rough parts of two objects catch one another as they rub together. The resistance that occurs depends on a value called the coefficient of friction. This is a measure of how much two objects interact with one another when rubbing together, and this depends on the materials that make up the two objects. The friction that takes place also depends on the mass of the moving object in the pull of gravity. All of these factors play a role in determining how much friction there is between two rubbing objects.

MATERIALS

- Two "two-by-four" or similar size wood planks, about 20 inches (0.5 m) long works best
- Small plastic tub with lid
- Sand or another dense material to fill the tub
- Water
- Freezer
- Books
- Ruler
- Protractor
- A carpeted area or a rug that can get damp (or a protective covering, such as a large plastic trash bag)

PREPARATION

- Make sure that the pieces of wood have surfaces that are similar in roughness and are about the same length and width.

- Take one piece of wood and soak it in water—this can be done by weighing the wood down in a bathtub filled with a few inches of water for at least 30 minutes.
- Place the soaked piece of wood in the freezer overnight.
- Fill the plastic tub with sand or another dense material and secure the lid.
- Use caution (you may want to wear gloves) when handling the wood—especially if it is rough—to avoid getting splinters.

PROCEDURE

- On the carpeted area or a rug, build a ramp using several books and the dry piece of wood. Make sure the same part of the wood leans on the books for all of the trials.
- Place the filled tub on the top of the ramp, with the edge of the tub lined up with the top end of the wood. Release the tub. If the tub slips down the ramp, remove the tub and lower the ramp height by removing a book. If the tub does not slip, remove the tub and raise the ramp height by adding a book. *How high is the stack of books when the tub slips? What is the height of the shortest stack needed for the tub to slip?*
- Using the protractor, measure the angle where the ramp meets the carpet when the ramp is set up using the shortest stack of books needed for the tub to slip. *What is the angle?*
- When you are done testing the dry piece of wood, retrieve the icy piece of wood from the freezer. (Because the icy piece of wood will thaw out over time, have everything in place and ready for testing it before you remove it.) *How does the icy wood feel compared with the dry wood?*

- Using the icy piece of wood, test it on your ramp with the tub just as you did using the dry piece of wood. *What is the height of shortest stack of books needed for the tub to slip on the icy piece of wood?* Again, use the protractor to measure the angle of the ramp. *What is the angle?*
- You can try testing each ramp with different book stack heights multiple times. *Are your results consistent, or is there some variation?*
- *Overall, which ramp—the one made with the icy or the dry wood—allowed the tub to slip at a smaller ramp angle?*
- **Tip:** If the icy wood thaws while you are testing it or seems dry, put it horizontally in the freezer and pour a little water all over the top of it. Leave it in the freezer for at least an hour. Test it again using the top side as the surface of your ramp.

SCIENCE FAIR IDEA

How easily would the tub slip on a ramp made of other materials? Make a similar ramp using a plank covered with sandpaper, oil, or aluminum foil. Alternatively, cover the bottom of the tub in these materials. *How easily does the tub slip using the different materials?*

EXTRA

If you know the ramp's height and length, you do not need a protractor to figure out its angle. All you need is to do some math: The equation used to solve the angle of the ramp is sin (theta) = opposite/hypotenuse. In the formula, "sin" stands for sine, "theta" is the angle, "opposite" is the ramp height and "hypotenuse" is the ramp length. *Does the calculated angle match the one you measured with your protractor?*

OBSERVATIONS AND RESULTS

In general, did the tub slip down the icy ramp better than the dry ramp? Was the smallest ramp angle needed for the tub to slip on the icy ramp smaller than the angle needed on the dry ramp?

When two objects rub together, there is always friction, and the surfaces of the two objects determine the amount that is generated. In this activity, you should have seen that there was less friction between the tub and the icy wood than from the dry plank. In other words, a smaller ramp angle was needed when using the icy wood than when using the dry wood for the tub to slip.

To calculate the force of friction, one must multiply the coefficient of friction by the force generated by the object, which was the mass of the tub in the pull of gravity. Although the mass of the tub and gravity did not change, the coefficient of friction of the two ramp systems did. The icy plank and the tub bottom had a lower coefficient of friction than the dry wood and the tub. You can easily feel this difference by carefully running your hand along the icy and then dry wood surfaces. The icy plank's frozen coating makes it much smoother than the dry wood.

CLEANUP

Discard the ramp and tub in the trash with the help of an adult to avoid unwanted splinters.

Speedy Science

How Does Acceleration Affect Distance?

IF YOU DROP A FEATHER AND THEN YOU DROP A ROCK, THEY WILL FALL AT DRASTICALLY DIFFERENT SPEEDS. HAVE YOU EVER WONDERED WHY THAT IS?

Have you ever dropped something and wondered how fast it was moving while falling? (If it was something fragile, you might not have been thinking about this at the time—you were likely too busy trying to grab the object!) We all know that gravity forces an object to fall. But how does this affect how quickly something falls and its impact? For example, did the object move faster right after leaving your hand, or just before hitting the ground? In this science activity, you will explore the relation between time and distance traveled when a moving object is under gravity's constant acceleration.

PROJECT TIME

about 20 minutes

KEY CONCEPTS

Physics
Velocity
Acceleration
Gravity

BACKGROUND

You know from experience that when you ride a bike downhill, it is easy to get going fast. Gravity is giving you an extra push, so you don't have to do all the work with the pedals. You also know from experience that the longer the hill, the faster you go. The longer you feel that push from gravity, the faster it makes you go. Finally, you know that the steeper the hill, the faster you go. The steepest "hill" you could imagine is not much like a hill at all, but rather a sheer vertical drop—where objects go into free fall and where gravity gives the biggest push of all. You wouldn't want to try that on your bicycle!

In free fall, gravity constantly accelerates an object (increases its velocity)—until it hits terminal velocity. Specifically, gravity increases a falling object's velocity by 9.8 meters per second (m/s) with every passing second. (Whereas velocity is measured in m/s, acceleration is measured in meters per second squared, or m/s^2.) How does this constant acceleration affect the distance that an object travels over time? In this experiment, you will roll a marble down a ramp to find out.

MATERIALS

- Long cardboard tube, such as an empty roll of wrapping paper, to make your ramp. It should be at least 2 1/2 feet (76 cm) long.
- A thin book or small wood block to raise one end of your ramp. It should be about 1/2 inch (1.2 cm) to 1 inch (2.5 cm) in thickness.
- Pair of scissors for cutting the cardboard tube
- Permanent marker
- Marble
- Timer. Make sure it can accurately count individual seconds. Many smartphones have a timer that is this accurate.
- A ruler (optional)
- A helper (optional)

PREPARATION

- Take a long cardboard tube and cut it straight along one of its (long) sides. Then cut it along the other side so that you end up with two long pieces that are each semicircles. You will use one of these pieces as the ramp for the marble.
- Take one of the semicircle pieces you just cut and raise one end slightly by placing it on a thin book or small block (no thicker than 1 inch [2.5 cm]—you want a low slope so that the marble does not roll too fast to measure).
- Use the permanent marker to mark a starting line across the high end of the ramp, about 1/2 inch (1.2 cm) from the end. You ramp is now ready for some marble-rolling action!

PROCEDURE

- Set the timer for one second and then hold a marble in place at the starting line.
- At the exact same time as you start the timer, release the marble (being careful not to give it a push as you let it go). At the same time, be ready with the marker to note the location of the marble after the second is up. If you have a helper, have them watch the marble for you. Use the permanent marker to mark where the marble was one second after releasing it. *How far did the marble travel?*
- Repeat this process at least nine more times. This means you should end up with at least 10 different marks on the ramp, showing where the marble was one second after releasing it each time. *Did the marble travel similar distances each time?*

- Repeat this process, but this time mark where the marble is at two seconds after releasing it. Do this at least 10 times, so that you have made at least 10 more marks on the ramp. *How does the distance the marble traveled after one second compare to the distance it traveled after two seconds? Specifically, how does the distance between the starting line and the one-second marks compare to the distance between the one-second marks and the two-second marks?*
- If you have a ruler, you can measure the distance the marble rolled each time from the starting line.
- *Overall, how did the distance the marble traveled change as it rolled down the ramp for a longer amount of time? What does this tell you about the effect of gravity on the velocity and acceleration of the marble as it rolled?*

EXTRA

Use a ruler to measure the marks you made on the ramp and calculate the average distance the marble traveled for one second and for two seconds (averaging the times separately). Then graph your results, putting the average distance traveled on one axis and the time on the other axis. *Based on your graph, how did the distance the marble traveled change as it rolled for a longer amount of time?*

EXTRA

Try repeating this activity using a longer ramp, such as a flat board that is at least 6 feet (1.8 m) long (with a groove down the middle, or another straight piece of wood glued along the length of the first board to act as a guide). Roll the marble down the longer ramp for increasing amounts of time (one second, two seconds, three seconds, etc.) until the time the marble goes past the end of the ramp. Mark the distance traveled on the ramp each time. *Using a longer ramp, how does gravity appear to affect the velocity and acceleration of a marble as it rolls downhill?*

EXTRA

Try repeating this activity with different balls, such as a rubber ball, large metal ball bearing, ping pong ball, etc. *How does the type of ball affect how it travels down the ramp?*

OBSERVATIONS AND RESULTS

Did the marble travel faster as it went farther down the slope?

When an object is in free fall, gravity increases its velocity by 9.8 m/s with every passing second. So after one second the object would be falling at a velocity of 9.8 m/s. After two seconds the object would be falling at a velocity of 19.6 m/s. After three seconds the object would be falling at a velocity of 29.4 m/s, and so on.

Although this activity was not performed in free fall (the presence of the ramp provides resistance to slow the marble down), the same idea applies here. Gravity should have constantly accelerated the marble as it rolled down the ramp in this activity. This can be seen by comparing the (average) distance between the starting line and the one-second marks to the distance between the one-second marks and the two-second marks. Specifically, the distance between the one- and two-second marks should have been greater than the distance between the starting line and the one-second marks, showing that the marble moved faster the longer it rolled. Depending on the exact conditions of your ramp setup, the distance between the one- and two-second marks may have been about 1.5 to 2.5 times the distance between the starting line and one-second marks.

CLEANUP

Recycle the paper scraps and return the marbles to their proper location.

Soaring Science

The Aerodynamics of Flying a Frisbee

FLYING A FRISBEE MAY SEEM LIKE THE FOCUS IS ON FLICKING YOUR WRIST, BUT THERE'S MUCH MORE HAPPENING THAN THAT!

Are you good at tossing a Frisbee? Have you ever wondered how a Frisbee is able to fly through the air so well? If you can throw a perfect, arcing curve right on target, you've already trained your arm to aid in the aerodynamics of Frisbee flight! In this activity, you'll investigate how the angle at which you throw the Frisbee affects its flight direction and distance. Next time you're out tossing a Frisbee, this little lesson in aerodynamics may help make your throws even more accurate!

PROJECT TIME

20 to 30 minutes

KEY CONCEPTS

Aerodynamics
Forces
Physics
Lift
Drag

BACKGROUND

Two key forces that act on a Frisbee during flight are lift and drag. Lift is the force that allows the Frisbee to stay airborne, and in flight it opposes the force of gravity on the disk's mass. The Frisbee itself creates this lift force as it flies through the air. Because of the Frisbee's curved shape, the airflow above it must travel at a higher velocity than that underneath, thereby creating low pressure above and high pressure below the disk. This pressure difference provides the lift. Drag is a resistant force on the Frisbee, perpendicular to the lift, and it acts against the disk's movement through the air. The angle at which the Frisbee is thrown, which we'll call the "launch angle" (aka the angle of attack), affects both lift and drag.

As a side note, you've probably noticed that a Frisbee doesn't travel far if it's thrown without spin. Spinning the Frisbee helps it fly by supplying angular momentum, which helps keep the Frisbee stable; the faster it spins, the more stable it should be.

MATERIALS

- A Frisbee
- Long string or hose
- Large open area in which to toss a Frisbee
- Tape measure
- A helper (optional)
- A piece of paper and a pen or pencil (optional)

PREPARATION

- Use the long string or hose to make a long, straight line in front of you, at least 24.5 feet (7.5 m) long. You will be throwing the Frisbee so that it is directed down this center line.
- Practice throwing the Frisbee down the straight line a few times so you get used to tossing it. If you have not thrown a Frisbee much before, you may want to try practicing it for a little while. **Tip:** A good way to throw a Frisbee is by standing sideways with the Frisbee held in front of you (near your non-tossing shoulder), then bringing the disk horizontally across you before you release it.

- If there is wind during any of your Frisbee throws, note the wind speed and direction.

PROCEDURE

- Throw the Frisbee as flat and horizontal as you can, aiming it down the center line you made. You can have a helper watch to confirm the angle at which you throw the Frisbee. *How far did it travel? How far did it veer from the center line, and in what direction (left or right)?* If you have a piece of paper and a pencil or pen, you can record this data and all following flight information.

- Throw the Frisbee as flat and horizontal as you can at least four more times. Each time throw the Frisbee with similar arm motion and speed, use a similar spin, and have the same release point. *How far did the Frisbee travel each time? How far did it travel away from the center line, and in what direction?*

- Throw the Frisbee tilted up a little, at a roughly 45-degree angle above the previous, flat throw. (You can imagine that if your arm were the hour hand of a clock, the previous throw would have been directed toward three o'clock and this throw should aim between one o'clock and two o'clock). Throw it this way at least five times. Other than changing the launch angle, try to keep all other aspects of the flights the same. *How far did the Frisbee travel each time when thrown at an upward angle? How far did it travel away from the center line, and in what direction?*

- Throw the Frisbee tilted down a little, aiming at an angle about 45 degrees below a flat throw (between four o'clock and five o'clock), at least five times. Again, try to keep all other aspects of each flight the same. *How far did the Frisbee travel each time when thrown at a downward angle? How far did it travel away from the center line, and in what direction?*

- *Did you see a consistent relationship between launch angle and flight direction? Is there a relationship between launch angle and distance? Based on aerodynamic principles, why do you think you saw these relationships?*

EXTRA

In this activity, you investigated how the Frisbee's launch angle affects its flight distance and direction, but you only tested a few angles. You can try this activity again, but test even more angles, such as angles in between the ones you tried in this activity. You can videotape your throws and then watch the video to analyze and confirm the angles at which you threw the Frisbee. *How well does the Frisbee fly using other launch angles? Is there an angle that consistently correlates with the "best" flight in terms of distance and stability?*

EXTRA

In this activity, there was not a focus on the effects of wind on a Frisbee's trajectory, but it can definitely be a factor. *How will the Frisbee's flight be affected by throwing it into the wind? What about across or with the wind? How does the launch angle change a flight in each of these conditions?*

EXTRA

You could compare the flight of a Frisbee with that of an Aerobie (flying ring disk). *What differences do you notice? Can you explain them in terms of aerodynamic forces?*

OBSERVATIONS AND RESULTS

When you threw the Frisbee as flat and horizontal as you could, did it generally fly relatively straight (not far to the left or right) and pretty far? When you threw the Frisbee tilted up a little, did it fly pretty high and far but not as straight, going off to the side near the end of its flight? When you threw the Frisbee tilted down a little, did it fly not nearly as far—and did it also go off to the side?

To fly well, the Frisbee needs enough lift, and not too much drag. When the Frisbee is thrown tilted downward, it does not have much lift and so it quickly falls to the ground. When the Frisbee is thrown relatively horizontal, it has a good amount of lift and consequently should fly relatively far—at least much farther than when the Frisbee was thrown tilted down. When an even larger launch angle is used, the Frisbee has more lift. You may have noticed, however, that although the Frisbee thrown upward flew relatively high, it probably stalled out rather abruptly near the end of its flight. This may have caused it to land gently and/or quickly go off to the side. Of the three launch angles tested In this activity, the horizontal launches probably resulted in the overall "best" Frisbee throws in terms of distance and straightness.

CLEANUP

Return the Frisbee to its proper location and discard the strings and paper.

Suspended Science

How Does a Hovercraft Hover?

HOVERCRAFTS MAY SEEM LIKE MAGICAL FLYING DEVICES, BUT YOU CAN DISCOVER HOW THEY WORK THROUGH A SCIENCE EXPERIMENT RIGHT AT HOME!

Have you ever ridden on a hovercraft? It is like gliding on a cushion of air! In fact, that's exactly what you're doing—a hovercraft is a vehicle that glides over a smooth surface on a cushion of air. Because a hovercraft can travel over flat land or water, it is an amphibious vehicle. In this activity, you'll get to build your own mini hovercraft using a CD or DVD, a pop-top lid from a plastic bottle, some glue, and a balloon. How will different amounts of air in the balloon affect how long the hovercraft hovers?

PROJECT TIME

about 30 minutes

KEY CONCEPTS

Vehicles
Air
Friction
Engineering
Aerodynamics

BACKGROUND

Because a hovercraft is a vehicle that glides over a smooth surface by riding on a cushion of air, it is also called an air-cushion vehicle, or ACV. How is the air cushion made, and how does it allow the vehicle to glide over it so freely? Vents or currents of slow-moving, low-pressure air are ejected downward against the surface below the hovercraft. Modern ACVs often have propellers on top that create the air currents. These currents are pushed beneath the vehicle with the use of fans. Surrounding the base of the ACV is a flexible skirt, also called the curtain, which traps the air currents, keeping them underneath the hovercraft. The trapped air currents can create an air cushion on any smooth surface, land or water! The cushion greatly reduces friction, allowing the vehicle to glide freely over the smooth surface below.

MATERIALS

- Pop-top lid from a plastic drinking bottle. (Reusable plastic drinking bottles sometimes use these kinds of lids.)
- An old CD or DVD that can be made unusable
- Craft glue or superglue
- A medium-size balloon (should be able to inflate up to at least 11 inches [28 cm])
- Stopwatch or timer
- Large flat surface for testing the hovercraft

PREPARATION

- Remove a pop-top lid from a plastic drinking bottle.
- Glue the base of the lid to the CD (or DVD) so that the lid covers the hole in the center. If you use Superglue, have an adult help, use caution, and follow all of the instructions and safety warnings on the packaging.
- Allow the glue to dry completely.

PROCEDURE

- Make sure the pop-top lid is closed.
- Blow the balloon up as large as you safely can without popping it, then pinch the neck so that no air can escape.
- Stretch the neck of the balloon over the pop-top lid, being careful not to let any air escape. Carefully center the balloon's opening above the pop-top lid opening. Your completed hovercraft should have the CD flat on the bottom, pop-tip lid above, and the inflated balloon's neck stretched snugly around the closed lid. Your hovercraft is now ready to do some hovering!
- **Tip:** If the pop-top lid detaches from the CD and you used craft glue to glue them together, carefully re-glue them using Superglue.
- Place the hovercraft on a flat surface. Start your stopwatch or timer, open the pop-top lid and push the hovercraft. Stop the stopwatch when the hovercraft stops hovering. *How long did the hovercraft hover?* Detach the balloon from the pop-top lid.
- Repeat this process two more times, inflating the balloon as large as you safely can, reattaching it to the pop-top lid, and timing how long the hovercraft hovers. *Did the CD hover for the same amount of time each flight, or was there some variation?*
- Repeat this process three more times, but this time only inflate the balloon to medium size. For example, if it took three breaths to blow the balloon up as large as you safely could, use only two breaths or a little less to inflate it this time. *How long did the CD hover when the balloon was only a medium size? Did the hovercraft hover for about the same amount of time for each of the three flights using a medium-size balloon?*

- Repeat this process three more times, but this time only blow the balloon up to a small size. For example, if it took three breaths to fill the balloon to maximum size, use only one breath or less to blow it up now. *How long did the CD hover using a small-size balloon? Did the craft hover for about the same amount of time each of the three times you tested it using a small balloon?*

- *Overall, which size balloon allowed the CD hovercraft to hover for the longest amount of time? Which allowed it to hover for the shortest duration? Why do you think this happened?*

EXTRA

In this activity, you made the balloon approximately different sizes. However, you could more accurately measure the different balloon sizes and see how this correlates to the time the hovercraft is suspended. Repeat this activity, but in this version, measure the circumference of the balloon each time you inflate it and estimate the volume of air inside by calculating the volume of a sphere. Alternatively, you can use either a water-displacement method: dunk the filled balloon in a large metric measuring container and determine the change in height of the water; or an air-displacement method: fill a large graduated cylinder with water and place it in a large tub of water, invert the cylinder underwater so that the liquid inside does not escape, release the air from the balloon into the cylinder opening so that the air travels up inside, and then look at how much air is trapped in the cylinder. *How does the volume of the balloon correlate with how long your CD craft hovers?*

EXTRA

Test the CD hovercraft on different surfaces. *On which type of surface does it work best?*

SCIENCE FAIR IDEA

Can you improve on this hovercraft design? Think about features of this craft that you can change, such as the construction materials, the lid's opening size, the hovercraft's shape or its size. *Could you add a skirt like that used on real ACVs to better contain the air cushion?* **Tip:** Foam board cut into different shapes may work well instead of a CD or DVD. *What modifications make the hovercraft stay suspended even longer? Can you use the action of a modified hovercraft for any practical purpose, or to solve a problem?*

OBSERVATIONS AND RESULTS

When using the large-size balloon did the hovercraft usually hover for the longest amount of time? When using the small balloon did the hovercraft usually hover for the shortest amount of time? Were the medium-size balloon's hovering times somewhere in between?

Although modern hovercraft often use propellers on their tops to create air currents that are pushed beneath it, in this activity, you used a balloon to create air currents. The air currents in your hovercraft traveled through the pop-top lid to go beneath the hovercraft. Because the pop-top lid stayed the same, the opening that allowed air to go beneath the hovercraft was always the same size. This allowed you to test balloons inflated to different sizes. The large balloon held the largest amount of air and should have allowed the hovercraft to stay aloft for the longest amount of time. However, some variation (of just a few seconds) among the three trials with each size may have been apparent because the balloon was probably not the exact same size each time you inflated it. Likewise, because the small balloon held the least amount of air and the medium balloon held an intermediate amount, the small balloon should have allowed the hovercraft to fly for the least amount of time and the medium balloon should have allowed it to hover for an intermediate duration.

CLEANUP

Throw away each of your used materials.

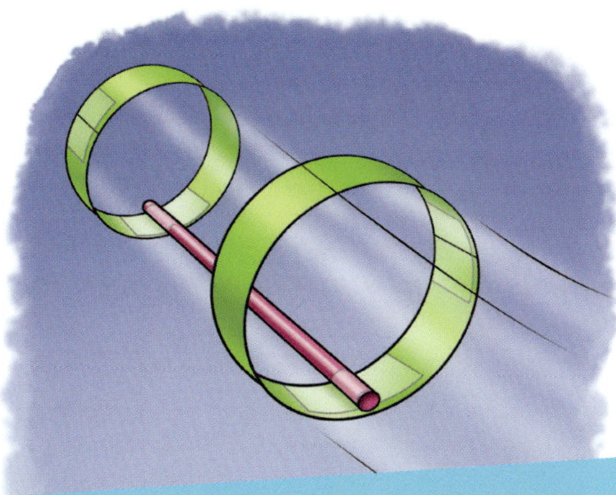

Loop-the-Loop with a Flying Hoopster

MOST PAPER AIRPLANES ARE FLAT, BUT THESE PAPER HOOPS CAN REALLY FLY!

Paper airplanes are fun to make and fly. Most designs resemble miniature planes—made of folded paper, with wings, stabilizers, and sometimes even flaps. These creations look like they are ready to soar. There are some designs, like the one you can try in this activity, however, that look so awkward one might imagine they would not fly at all. Find out if it will really soar!

PROJECT TIME

20 to 30 minutes

KEY CONCEPTS

Gravity
Thrust
Lift
Drag

BACKGROUND

Although there are many different paper airplane designs, almost all of them have a flat winglike structure, which, like a traditional plane, helps create lift to keep the plane in the air. The plane in this activity, which is reminiscent of Phillip Swift's design entered in *Scientific American's* International Paper Airplane Competition in 1967, gets the lift needed to glide from two rings, instead of wings. Normal wings have wing tips that generate vortices, adding drag (which isn't ideal for an object trying to stay aloft). The lift-providing rings have no wing tips and so they have less drag. The streamlined shape of the "hoopster" also makes it easy to throw accurately.

MATERIALS

- Scissors
- Ruler
- Pen or pencil (optional, for measuring out paper to cut)
- 3- x 5-inch (7- x 12-cm) index card (or a file folder or some other stiff paper)
- Piece of printer paper (or similar stiff paper)
- Clear plastic tape
- Stiff plastic straw (that does not have a bendy top)
- Measuring tape (or long string)
- Open space—indoors or out—where you can throw a paper airplane and measure the distance traveled. A long hallway or large room might work best indoors. Outdoors, try to find a time when and area where it is not too windy.

PREPARATION

- Use the ruler (and pen or pencil) to measure 1 inch (2.5 cm) increments along the short side of your index card.

- Cut a file card the long way into three equal strips, generating three strips that are 1 inch (2.5 cm) wide and 5 inches (12 cm) long. (If you're using stiff paper, use a ruler to make three strips that size.)

- Place one piece of tape on the short end of one strip. Curl the paper into a full loop and tape the two short ends together.

- Place the other two strips end to end, so they overlap a little. Tape them together to make one long strip.

- Place another piece of tape on one end. Curl the strip into a larger loop and tape the ends together.

- Place one end of a straw onto the middle of a strip of tape. Place the big loop on top of the straw and fold the tape up the sides of the loop.

- This part can be a little tricky: Place another strip of tape at the other end of the straw and press the small loop very gently onto the tape. Move it around until it lines up with the big loop then press the tape down firmly. *Does this look like any aircraft that you have seen before? How well do you think it will glide? Why?*

- Use the other sheet of paper to construct your favorite standard paper airplane.

PROCEDURE

- Get ready to fly your "hoopster!" Hold the hoopster in the middle of the straw, with the small "hoop" in the front. Throw it like a spear. (It may take a little practice to get the throwing technique right). *How did it glide?*

- Designate a "starting line" from where you can throw both planes.

- Throw the hoopster. Measure from the starting line to where it landed. *How far did it go?*
- Repeat this at least two more times. *What is the average or the median distance it traveled? Do you think it will beat the paper airplane?*
- Practice throwing your standard paper airplane. *How did it glide?*
- Now stand at your same starting line and throw the standard paper airplane. Measure from the starting line to where it landed. *How far did it travel?*
- Repeat at least two more times with the standard paper airplane. *What is the average or the median distance it traveled?*
- *Which paper glider soared farthest? Were their flight patterns different? Did you need to throw them differently?*

EXTRA

Try changing the dynamics of each plane with an addition. Put a paper clip at the bottom of the small hoop. *What happens when your hoopster has cargo? If you add the same cargo to the standard airplane, does it have a similar effect?*

EXTRA

Make a really long hoopster with two straws. Does it fly farther than the short one? To do this, cut a little slit at the end of one straw and pinch it so it fits inside the other straw, then tape them together.

EXTRA

Make a double hoopster. Use two index cards to create two little hoops side by side on one end and two big hoops side by side on the other. *Do the extra hoops make a difference in how it flies?*

OBSERVATIONS AND RESULTS

The same principles are at work in all fliers. Gravity is pulling them down but at the same time air helps hold them up. If you drop a flat sheet of paper, it flutters to the ground, wobbling this way and that. To reach the ground, the paper has to push the air aside. Each time the paper wobbles it lets more air flow past it and it gets a little closer to the ground. Because a flat sheet of paper has to push a lot of air out of the way, it falls slowly. Crumple that same piece of paper into a ball and it will fall straight down without fluttering. The ball of paper has the same weight as before but air can flow around it easily and it falls quickly.

For something to glide through the air, there must be a balance between the gravity that pulls the glider down and the lift produced as it glides through the air. A design must balance these forces so that the glider moves forward, propelled by the pull of gravity.

What works well? What doesn't work at all? Another part of fooling around scientifically is keeping track of your results. Keep track of alterations that you try. If you come up with a new design that you like, share it!

CLEANUP

Discard used materials in the recycling or trash, and return other materials to their proper place.

THE SCIENTIFIC METHOD

The scientific method helps scientists—and students—gather facts to prove whether an idea is true. Using this method, scientists come up with ideas and then test those ideas by observing facts and drawing conclusions. You can use the scientific method to develop and test your own ideas.

Question: What do you want to learn? What problem needs to be solved? Be as specific as possible.

Research: Learn more about your topic and refine your question.

Hypothesis: Form an educated guess about what you think will answer your question. This allows you to make a prediction you can test.

Experiment: Create a test to learn if your hypothesis is correct. Limit the number of variables, or elements of the experiment that could change.

Analysis: Record your observations about the progress and results of your experiment. Then analyze your data to understand what it means.

Conclusion: Review all your data. Did the results of the experiment match the prediction? If so, your hypothesis was correct. If not, your hypothesis may need to be changed.

GLOSSARY

acceleration: Increase in the rate or speed of something.
aerodynamics: The science that studies air movement and how objects move through air.
centripetal force: The force that draws an object moving in a circle toward the center of the circle.
engineering: The use of science and math to build better objects.
hypotenuse: The longest side of a right triangle, opposite the right angle.
force: Strength or power placed on an object.
mass: The amount of matter in an object.
motion: Movement.
orbit: To travel in a circle or oval around something, or the path used to make that trip.
phenomenon: Something that can be studied and that may be unusual or difficult to understand or explain fully. The plural is phenomena.
physics: The branch of science concerned with the nature and properties of matter and energy.
pressure: A force that pushes on something else.
satellite: An object that circles Earth in order to collect and send information or aid in communication.
symmetry: Made up of exactly similar parts facing each other or around an axis.
velocity: The speed of an object in a given direction.
wind power: The process by which the wind is used to generate mechanical power or electricity.